Existence of
Design Codes
in
Living Organisms

EXISTENCE OF
DESIGN CODES
IN
LIVING ORGANISMS

MIRABOTALIB KAZEMIE

authorHOUSE®

AuthorHouse™
1663 Liberty Drive
Bloomington, IN 47403
www.authorhouse.com
Phone: 1-800-839-8640

Published by AuthorHouse   06/14/2013

ISBN: 978-1-4817-5384-5 (sc)
ISBN: 978-1-4817-5383-8 (hc)
ISBN: 978-1-4817-5385-2 (e)

Library of Congress Control Number: 2013909071

Any people depicted in stock imagery provided by Thinkstock are models, and such images are being used for illustrative purposes only.
Certain stock imagery © *Thinkstock.*

This book is printed on acid-free paper.

Because of the dynamic nature of the Internet, any web addresses or links contained in this book may have changed since publication and may no longer be valid. The views expressed in this work are solely those of the author and do not necessarily reflect the views of the publisher, and the publisher hereby disclaims any responsibility for them.

CONTENTS

INTRODUCTION

We have a fundamental question in biology that is still waiting for the right answer. It is about the molecular mechanism that determines the designs of biological organisms. How is it possible that many living organisms despite sharing similar regulative systems during their embryonic development and having similar structural proteins differ from each other morphologically, physiologically and behaviorally?

To provide answer to this question, I present in this book a comprehensive biological theory that I believe to be able to show that the design, function, and behavior of a biological organism are determined by a design code that is encoded by some invariant proteins, so that every biological organism has its own specific design code. The proteins that are part of the design codes, unlike many other proteins, cannot tolerate mutations.

In the light of this theory the issues of adaptability and variability of the living organisms will be better understood.

They are defined as innate potentials in living organisms that serve not only their survival, but also the integrity of their identities. The theory of design code takes also a reasonable approach to help differentiate between the mechanisms that are involved in development and those that are in charge of biological diversity. I assume that the design code proteins have the supreme kind of intelligence, the basis of cellular intelligence, and paramount for the development of instincts, learning faculties, and behaviour. Alteration of any of the proteins involved in a design code leads to the destabilization of the code. This can occur in an egg cell, in a cell in the very early stage of embryonic development, and in a somatic cell. While the first two situations can be lethal, or cause monstrosities, destabilization of a design code in a somatic cell can turn that cell into a cancer cell.

Design Codes
preset the Designs
of Living Organisms

Regarding the origin of living organisms on earth, I assume that prior to the emergence of the first living cells there had been protein molecules that worked together to built such cells around themselves. Those proteins were invariant and continue to remain invariant since then. I call them 'primordial proteins'. The first cells that they built I name 'urprimordial cells'. They were able to grow, reproduce and communicate with the environment surrounding them. Subsequently, inside the urprimordial cells, the primordial proteins dynamically had been changing the pattern of their interactions, producing design codes to encode the designs for numerous organisms to emerge afterward. An urprimordial cell in which the encoding process takes place turns into a 'primordial cell'. The encoding process itself I call 'primordialization'. In this way a primordial cell obtains the information for the ultimate appearance, functions, and behavior of the organism to which it develops.

The number of the invariant proteins that were working together to build the design codes constituted only a fraction of the urprimordial cells' proteins, but they could theoretically encode thousands of different design codes, depending on their number, their location inside or on the surface of the cells, and their manner of interaction with each other. In other words, design codes differ from each other by the way those invariant proteins are combined and arranged during the encoding process. Today I consider the following invariant proteins as the members of primordial proteins and the likely candidates to encode organisms' designs: growth factors, growth-factor receptors, proteins that are involved in signal transmission, proteins such as those that determine the polarity of organisms, proteins in charge of symmetries in developing embryos, and proteins that are in control of DNA molecules.

In a multicellular organism the design code is translated into a comprehensive plan that has to be read after each cell division throughout the entire processes of development and differentiation. In other words, each cell in a developing embryo has to be provided with a set of instructions defining its specific duties. How this takes place? I assume that at any level of development and cell differentiation the organism's design code regulates the activities of the subordinate regulative proteins, whose jobs are to lead the cells and organs to develop within the preset boundary of the organism's design. Passing of information in this manner will continue from one cell generation to the other until the structure of the organism is complete.

In metazoa, the egg cells must contain the complete set of the invariant proteins that are required for encoding the ultimate designs of the organisms, because no mRNA synthesis occurs before the first cleavage, which starts some hours after fertilization.

Encoding deigns for new organisms through reprogramming of existing design codes

While in primordial times the design codes were made inside the urprimordial cells, these cells have no longer been necessary for this process, because designs for new organisms can be encoded inside any cell of any living organism through reprogramming of an existing design code. I call this process 'reprimordialisation'. To generate a new form the existing organism or parts of it turns into undifferentiated cells, by which the design code proteins in those cells rearrange themselves into a new design code, while the amino acid sequences of all participating proteins remain unaltered.

The new organisms encoded in this way can follow different paths, depending from their designs. There are those that are coexisting with the original organisms, or others that live in a cycle with the original ones, or those that continue their own independent existence. I describe, in the following, examples of the organisms that are emerging through reprogramming of design codes. It needs to be mentioned that despite their uniqueness these new designed organisms have been misinterpreted and mixed with 'differentiation', or called 'induction', 'metamorphosis', or 'life cycles'.

1. I consider placenta as an organism, not as an organ. If it were an organ, then it must be integrated into the anatomy of the organism that continues to grow after the birth. Furthermore, the fact that it nourishes inside the uterus independently from the foetus also underlines this notion. Since it is an organism, now it needs to be explained how two organisms can develop from the same egg cell. As we know some of the external cells of the placental embryos at 8-16-cell stage (the merula) develop into a trophoblast, leading to the formation of placenta. I assume that the external merula cells in order to be able to lead to the formation of placenta, first of all, the original design code in those cells must be reprogrammed into a new design code, which then determines the design of the placenta. It needs to be mentioned that the formation of placenta should not be mixed with the ability of embryonic cells in some organisms, such as the larvae of Trematoda, to reproduce offspring, whereby the daughter embryos and the embryo that gives rise to them have the same design code.

2. Reprogramming of design codes can also take place by partial or total disintegration of some organisms under a process that has been called, incorrectly, metamorphosis. Let me explain this by taking Drosophila as an example. The fly, contrary to the current notion, is not the product of further development of the larva, because it has its own distinct design. The design of the fly seems to be encoded

inside the imaginal disks, and the design of the larva inside the syncytium [A fertilized egg begins its development with a series of very rapid nuclear divisions with no cell division. Most of those nuclei migrate from the middle of the egg toward the surface, where they form a syncytium. Later plasma membranes grow inward from the egg surface to enclose each nucleus, converting the syncytium into a blastoderm consisting of some 5000 separate cells. Thereafter, the larva emerges and passes through three stages, or instars, separated by molts in which it sheds its old coat and cuticle and lays down a larger one. When the life of the larva nears the end, its tissues break up. Its disintegrated epidermis cells are deposited in different segments of that organism (imaginal discs)]. I assume, that during disintegration of the epidermis the proteins building the design code of the larva breaks apart and then inside the imaginal disks reassemble in a new fashion, building the design code of the fly. At syncytium, in turn, the design code of the fly dissolves and its proteins combine in a renewed fashion to encode the design of the larva.

3. The emergence of a new form through reprogramming of a an existing design code is also very obvious by nauplius larva of parasitic copepod *Haemocera danae*, when it becomes parasite in a polychaete worm. Inside the host the life of nauplius larva reaches an end; its structures disintegrate, leaving no sign of a larva behind. What is left after this disintegration is an ovoid mass of small undifferentiated cells. These cells then give rise to a new form, free-swimming

organisms inside the worm.

In this place, by bringing further examples I want to substantiate my notion that larvae, no matter to which taxonomic group they belong, are distinct biological forms. They are active, feed themselves, and with some exceptions they live independently. In addition, larvae and the 'adult' organisms can live under very different environmental conditions. In case of *Haemocera danae,* the larva can live free, while the 'adult' organism is a parasite. On the other hand, a larva can live parasitic, while the 'adult' organism lives free (e.g., Gordioidea). Finally, many larvas instead of turning into the 'adult' organisms, can reproduce independently, either asexually (e.g., Cecidomya), or sexually (Cetenophora).

4. A reprogramming of design code must take place when parts of a hydroid polyp disintegrate and lead to emergence of medusas. The polyp usually produces its own kind, i.e. other polyps, through budding. Occasionally, from some of those buds medusas emerge instead. This occurs when the buds disintegrate and form gametangia. It is during this process that the polyp's design code is being replaced by a new design code, specific for the design of the medusa.

5. A reprogramming of design code is also involved in turning the unicellular amoeba *Dictyostelium discoideum* into a multicellular slug. When facing starvation, the amoebae aggregate into mounts, whereby the amoebae cells turn into undifferentiated cells. It seems that in these cells the design

code of amoebae changes into a new design code and give rise to the emergenge of multicellular slug.

Design codes modulate the fuctions between unicellular and multicellular organisms

Functions in multicellular organisms arise from redesigning of the functions that the urprimordial cells were already equipped with, in such a way to fit the needs of the designed organism. To show that this was possible, one needs to think about the designing of slug inside the mount of unicellular amoebae, where the whole functions of the multicellular slug derive from the redesigning of the functions of the unicellular amoebae. The slug has light sensitive structures at the anterior end of its body. These light sensitive structures are the result of rearrangements of the light sensitive molecules and structures present in those single undifferentiated cells of the mounts. With regard to the development of sight in invertebrates and vertebrates, the principle is the same. The eyes of these organisms differ from each other, because they have been designed differently from the similar light sensitive cells in urprimordial cells. In none of those organisms it can be said that they and their eyes and other functions are 'evolved' stepwise from each other.

According to the design code theory no biological form can be transformed into another, because those proteins cannot tolerate alterations. When any of the proteins involved in the design code of a cell is altered genetically, this would have disastrous consequences for that cell or even for the organism as a whole. In case such an alteration takes place in an egg cell, that cell would either die or give rise to development of a non-viable body, a deformation. For example, an organism with an anencephaly caused by mutation of a design code protein(s) is a deformed body. Very different from this kind of anencephaly is the anencephaly in chelicerates, which are so designed that they have no head and the organs which are operating in other organisms in head section in these are located in their prosoma section [the bodies of chelicerates are subdivided into two sections, the prosoma (front) and the opisthosoma (back).

When alteration of a protein involved in a design code occurs in a somatic cell, that cell would turn into a cancer cell. I will talk about this under the heading 'Destabilization of a cell's design code leads to initiation of cancer' later. Persistent to the design code theory, I want to emphasis that the term *transformation* cannot be applied to cancer; instead, I call cancer an *aberration*.

Design Codes
require that all
Regulative Proteins
work coherently

In all phases of development in animals and plants, it is thanks to the intelligence of their proteins that differentiation and proper tissue and organ formation proceed according to the instructions encoded in their design codes. The intelligence of the surface adhesion proteins, for example, enables the differentiated cells to assemble with their kinds. In experiments where cells from different parts of an early amphibian embryo were artificially dissociated and put into a random mixture, it has been shown that the cells in the mixture were still able to sort out according to their origin. Intelligent cooperation among protein molecules on the surface of a living cell is needed to respond to environmental changes successfully. Indeed, a cell is able to receive different kinds of chemical and physical signals from the environment and convey them from its surface along different pathways to several locations inside its body. Bacteria such as *E. coli*, for example, has been shown to sense chemical attractants over a concentration range of several orders of magnitude and respond to extremely low concentrations of nutrients. This is possible, because proteins that act as receptors on their surface can communicate with each other and make jointly an estimate of the concentration of the surrounding molecules.

Protein intelligence is also behind the functioning of biological clocks. These are self-sustaining clocks that give the rhythm inside the living organisms, whether unicellular or multicellular. One biological clock, which is identifiable in all living organisms, is the one that determines the sequence of the events during cell division. Though the duration of cell division may vary from a few minutes to many hours, depending from the organism, the sequence of the events during this process is, however, timely fixed. The presence of biological clocks in all living cells is evidence for the fact that, since the emergence of urprimordial cells, protein molecules inside the cells interact with each other in an ordered fashion.

Intelligence of protein molecules can very clearly be discerned from their ability to manipulate the DNA molecules. As we know, inside the cells there are numerous proteins that are in charge of DNA synthesis, its storage, repair, modification, and utilization as blueprint for protein synthesis. Proteins are involved in preventing entry of a damaged DNA molecule into the S and M phases of the cell cycle until the damage is repaired. Blocking entry into the S phase prevents the replication of a damaged DNA molecule, which otherwise would lead to defect daughter strands. In addition to the DNA-damage-repairing proteins, cells have also protein molecules that are capable of changing the sequence of a DNA molecule by adding new nucleotide sequences to it, or cutting off sections of variable lengths from it. They can remove parts of DNA sequences known as mobile elements from one location of the molecule and insert them into another location of it. [Mobile elements are DNA sequences that range in size from several hundreds to many thousands of base pairs]. By using these DNA elements, organisms have a mechanism available to change the structural architecture of its DNA blueprint. In the single-celled organisms called hypotrichous ciliates dramatic rearrangement of their DNA had been observed. These organisms in addition to their main nucleus have one or more micronuclei. During sexual reproduction the micronuclei from the two partners fuse and give rise to new micronuclei and a macronucleus. As the macronucleus takes shape, not only are the DNA segments between the coding regions removed, but the coding regions put back into correct order, all in a matter of hours. Furthermore, due to the intelligence of protein molecules, the events inside the cell can be reprogrammed. For example, in experiments where the nucleus of a differentiated cell is transplanted into the enucleated Xenopus egg, the nucleus became reprogrammed to resemble that of the normal egg nucleus [A typical amphibian egg is so large that using a fine glass pipette, one can readily inject into it a nucleus taken from another cell. A complete swimming tadpole could be produced, for example, from an egg whose own nucleus, which has been destroyed by ultraviolet radiation, was replaced by a nucleus from a frog red blood cell].

Proteins were found to be able to give themselves the folding they want. As studies on a number of proteins have shown they have intramolecular amino acid sequences that are in charge of this activity. Furthermore, studies on a protein called *split intein* have shown that this protein is capable of bringing together two pieces of different proteins, knitting them together, and then neatly cutting itself out. Another example of engineering abilities of cell proteins is the RNA interference system or post-transcriptional gene silencing. In this system an enzyme produces small RNA molecules called small interfering RNAs by cutting long double-stranded RNA into pieces of about 21 nucleotides long. Then one strand of each small RNA piece is loaded onto a protein called *Argonaute* protein, generating a RNA-protein complex that binds to viral targets by base pairing and cuts it.

Design codes use DNA as blueprint

At the time when the first living cells emerged on earth primordial proteins synthesized DNA as one of the constituent parts of the primordial cells, designed to be used as blueprint for their own primary structures and that of all other cell proteins, and at the same time as a mean of transferring information from one generation of cells to the other. The primordial proteins by knowing that a molecule to fulfil those two requirements is susceptible to environmental factors, had to design at the same time measures for its safeguard. In the following, I give examples of such measures.

Ahead of cell division the DNA replicates are already condensed and packed in a non-functional state. The cell proteins take precautious measures to prevent a damaged DNA molecule during cell cycle from entry into the S phase and M phase until the damage is repaired. Blocking entry into the S phase prevents the replication of the damaged

DNA, which otherwise could cause change in the daughter strands synthesized on damaged templates. Equally important is a surveillance system known as the spindle checkpoint, that recognizes the presence of chromosomes whose kinetochores (spindle attachment sites) are not properly attached to the spindle. This system ensures that chromosomes during the process of cell division segregate accurately by inhibiting chromosome segregation until chromatin separation is complete.

The very rapid cycles of DNA replication at the beginning of development hinder transcription. As it has been mentioned earlier the first cleavage occurs 17-20 hours after fertilization, and no mRNA synthesis takes place during this period. This means that the whole process that leads to the first cell division is controlled by the proteins already present inside the zygote. This is important, because it makes sure that no altered copy of any design code protein is produced, which could replace the proper ones inadvertently. Alterations of DNA sequences can occur not only during cell division and development, but also at any time in any cell of a multicelluler organism. They can be caused by environmental factors such as radiation, mutagenic chemicals, etc. Since alterations in the sequence of DNA molecules can change the amino acid sequences of proteins, they can impair the function of any cell protein if left unchecked. In cases where the mutated proteins belong to somatic cells, they can lead to disruption of normal cell functions. In situations when DNA mutations happen in germ cells the matter gets worse. Here it is not one individual that will be affected, but all members of the offspring that inherit the mutated DNA sequences will be suffering from the consequences, ranging from malfunctioning and deformation to death. The later is certain when the mutations of DNA affect proteins that are part of the organisms' design codes.

Organisms, in addition to what are mentioned above, are equipped with measures to prevent or in cases when it happens correct the damage. The measures include checking the inaccuracies in their DNA blueprints, and DNA repair mechanism. In cases when the repair system cannot deal with the altered DNA, the organism can then try to inhibit the division of the cells that are carrying mutated DNA sequences.

The means by which organisms can deal with DNA mutations do not end here. Living organisms have also other ways at their disposal to overcome the problem, such as the RNA silencing mechanism, which silences unwanted DNA elements. It needs to be emphasized that living organisms not only have the ability to correct the sequences of their DNA molecules that are altered by the environmental factors, but also the ability to alter it autonomously. One form of DNA manipulation is DNA rearrangement using mobile DNA elements. This includes nucleotide substitutions, large and small deletions, large and small insertions, and translocations of DNA sequences. Some organisms can even change the sequences of their preformed nucleic acids, a process known as RNA editing, common in chloroplast and mitochondria of fungi and protozoa. In other organisms it has been shown that they can produce new sequences, even at a very fast rate. The best known example for this is the production of DNA blueprints for the variable regions of immunoglobulin molecule in vertebrates as part of their precaution to respond to the enormous number of antigens, most of them had never been encountered by the former generations of the organisms in the past [The rate of changes in these sequences has been estimated to be 10^{-3} per nucleotide pair per cell generation]. This is in conformity with the design code theory, because they produce new antibody proteins to maintain their identities; they want to survive as distinct forms.

The high content of non-functional DNA in multicellular organisms is possibly the result of dumping non-coding DNA sequences

The existence of large piles of non-functional DNA in different living organisms is in agreement with the design code theory. In human with a DNA content of 3.2 picograms per cell, only a small fraction of its DNA codes for proteins—less than one percent. Taking the DNA content of an human cell as reference, the DNA content per cell of wheat, broad beans, and garden anions (7.0, 14.6, and 16.8 picograms, respectively) ranges from about two to more than five times as that. Tulips have ten times as much DNA per cell as humans. The cells of a lungfish have a DNA content 17 times higher than that of human cells. Sixty percent of these DNA piles seem to be non-functional; half of them are cytologically detectable as heterochromatin, and the other half consists of highly repetitive sequences.

It is important to find out the reasons behind the accumulation of large amounts of non-functional DNA in multicellular organisms. Based on the qualities of DNA described before, the following point can be made: organisms cannot afford to become incapacitated or die each time when a harmful DNA mutation alters the sequence of any of the proteins that is part of their design code. To prevent this from happening, they first try to use their DNA repairing system to correct the damaged DNA sequence as far as possible; when this is not sufficient, the other possibility is through excisions and insertions of nucleotide sequences with the help of mobile elements, on a trial-and-error basis, until the original sequence is restored. At the end, when the organism succeeds in restoring the DNA sequence into its original state, it has to dispose itself of those faulty DNA sequences that were produced during the correction process. One way of doing this would be through breaking it back into nucleotides; but this may cause for the cell more problems if inadvertently the enzymes break its normal DNA sequences. A better way of getting rid of the faulty DNA sequences seems to be their accumulation. I assume that this process, which I refer to as dumping of non-functional DNA sequences, is happening routinely inside the cells of multicellular organisms. In the examples given above, the high content of lungfish DNA could be the result of 400 million years of damping altered DNA sequences in the cells of this animal. Damping of non-functional DNA sequences can also be inferred from the observations that point to the existence of intra-group variations of DNA content. The amount of DNA inside insect groups, for example, varies by a factor of hundred. Similar variations have been detected among amphibians. A cell of a bullfrog contains twice as much DNA as that of a toad.

DESIGN CODE AND DEVELOPMENT

According to the design code theory the structures and shapes of living organisms at all phases of their development are determined by their design codes. However, the supporters of the evolution theory are seeing this as evidence that living organisms are evolved from common ancestors, which is an erroneous conclusion by them. The reason for the structural similarities of developing multicellular organisms is because of the use of similar structural elements by their design codes as basic axis of orientation for the organs to develop. In the following, I describe the interpretations of the supporters of the evolution theory about some of those structures and my own evaluations on the basis of my theory.

According to the evolution theory the fact that biological organisms share similar structural features during their development is an indication that many structures in today's living organisms are representing the transformed patterns of the same structures that originally had been existed in their remote ancestors. For instance, from observation that embryos of mammals, birds, and reptiles in their earliest stages are looking similar they assume that they arise from the same ancestors. Furthermore, they believe that the embryonic and larval stages are more or less completely the condition of the progenitor of a whole group of animals in its adult state. In other words, if two or more groups of animals, no matter how much they may differ from each other in structure and habit in their adult conditions, pass through closely similar embryonic stages, they are closely related by being descendants of the same ancestors. Referring to the embryos of mammals, birds, and reptiles, they assume that they are the modified descendants of some ancient progenitor which was furnished in the adult state with bronchi, a swim bladder, four fin-like limbs, and a long tail, all fitted for an aquatic life. In the following I re-examine the reasons for the similarities of organs in the light of the design code theory.

From the examples mentioned above, I examine first the case of four fin-like limbs, which the land vertebrates' embryos supposedly had. During early development of land vertebrate' embryos what emerges first in sites of the future limbs are small tongue-shaped buds, four identical of them. These buds show no similarity with fins or other forms of extremities that appear at their position at a later stage of development. With regard to the composition of their cells the two pairs of limb buds seem to show no difference. They are undifferentiated at first, showing no hint of the subsequent skeletal patterns. Their only differences are their location along the body axis. It is only from their position along the body axis that one pair develops into forelimbs or wings and the other one into hind-limbs.

Next, among the structures that have been considered for very long as the markers of phylogenetic kinship between mammals and the jawless fish are the seven cartilaginous gill arches of the adult jawless fish that form the opening through which water flows to provide oxygen for the capillaries of the gills themselves. These structures are also present in developing mammals' embryos, but later are transformed into different structures in adults. The first arch develops into the two bones of the inner ear, the hammer and anvil; the second one develops into stirrup. The rest of them develop into various cartilaginous structures, like thyroid cartilage and trachea cartilage. The idea of evolution of cartilaginous arches in mammals' embryos from the same structures of jawless fish cannot be true. If it were the case that during the past hundreds of million years those structures have undergone such changes that they ultimately turned into different organs, then one could expect that the original structures should have been disappeared from all phases of development in all vertebrates, not that the same structures still linger in their original form in these animals.

Another organ that has been considered as a phylogenetic marker, backed by evolution theory, is the notochord. It builds the longitudinal axis of the invertebrates called cephalochordates [Noticeably a similar organ is present during the embryonic development of vertebrates]. According to fossil records these invertebrates have been living on earth much longer than vertebrates; the members of their living representative called Amphioxus live still unchanged in many waters today. The finding that their notochord resembles a similar structure in vertebrates' embryo led some evolutionary biologists to believe that these invertebrates are the ancestors of all vertebrates. This belief is unfounded. According to the design code theory the reason for the presence of notochord in both Amphioxus and vertebrates is explainable: formation of notochord is encoded in the design codes of cephalochordates and vertebrates as well, with one difference that in cephalochordates it is supposed to stay beyond their embryonic period of development and in vertebrates only during that period. They all need a notochord structure for the reason that organs do not develop all at once. Their proper sequence and positioning within the developing embryo require a body axis for orientation. In developing embryos in vertebrates, notochord precedes the formation of spinal cord, thus providing the embryo with the essential longitudinal body axis. It is involved spatiotemporally in the formation of many other organs as diverse as brain, heart, kidney, segments of muscles, etc. Without the notochord a vertebrate embryo may not be able to develop its nervous system, vertebrae and many other organs, properly. Furthermore, the same argument that I brought in connection with cartilaginous gill arches is also valid with regard to notochord. If spinal cord in vertebrates were evolved from the notochord of cephalochordates, then the notochord should have been replaced completely in vertebrates. I want to reinterpret two other observations. The first one concerns the ability of the alimentary canals in the larva to a dragon-fly and in the fish Cobites to respire, digest and excrete. The second one is with regard to a change of function in the Hydra, when the animal turns inside out, so that the exterior surface digest and the stomach respire. The abilities of the alimentary canals in the larva to a dragon-fly and in the fish Cobites to respire, digest and excrete were considered by supporters of evolution theory as indications for evolution of one organ into another. On the contrary, according to design code theory embryonic cells during the development of an organism undergo a pattern of differentiation that fits the design of that form. The organ-specific and other local subordinate regulative systems are bound to strictly follow and carry out the instructions of the design code. When it is required that some cells be equipped so to carry three different functions of respiration, digestion and excretion at the same time, the subordinate regulative systems have to carry out this plan during differentiation accordingly. Concerning the change of function in Hydra, it is possible that according to the design of that organism their cells are programmed so to be able to shift their functions. I want to elaborate further on this issue focussing on the mutations that displace body structures in Drosophila fly. In this insect the decision as to whether an antenna or leg should develop in its head region depends from the presence or absence, respectively, of a protein known as *Antennopedia* protein [This protein belongs to the group of homeotic selector proteins. These proteins maintain the distinction between body segments; their mutations change structures in one body region into structures appropriate to other body regions]. The observation that in the absence of this protein leg sprouts in the head instead of antenna has been taken as evidence that insects are evolved from multipeds. I assume that this is not correct. If antennae were evolved from legs located in the same positions, then the subordinate regulative system for making legs in these locations in multipeds should not continue to exist in this fly; it should rather been replaced completely by the subordinate regulative system that makes antennae. According to the design code theory the reason for this phenomenon is at hand. Apparently, the cells at the antennae regions of the head of this fly have the potential to develop in both directions. Normally, the design code of Drosophila instructs the subordinate regulative system in charge of building antenna to build and put antennae in the specified spots of the head. However, in the absence of the latter system leg subordinate regulative system put automatically legs in those spots of the head. In other words, a loss of function of proteins responsible for making antenna triggers automatically proteins in charge of making legs to develop legs, instead. The same is true regarding the phenomenon observed due to the loss of function of other homeotic proteins. For instance, in segments belonging to the thorax region abdominal organs develop, and reversely in segments belonging to the abdominal region thoracic organs develop. This phenomenon does not indicate that thoracic organs are evolved from organs that were abdominal ones in the past, or conversely.

DESTABILIZATION OF A CELL'S DESIGN CODE LEADS TO INITIATION OF CANCER

C ancer cells share many characteristics such as their potential for indefinite growth, the increased mobility of their cell membrane and membrane proteins, their diminished ability to adhere to a solid surface, and their high growth activity even without attachment to cellular matrix. For the reason I describe below, I consider them as cells with 'defective design codes'.

Normal cells, both embryonic and differentiated, become cancerous when one or more of some important cell proteins, such as growth factors, growth-factor receptors, protein kinases, and nuclear proteins are altered. It is for this reason that the normal genes for those proteins are called proto-oncogenes. Normally, one would expect that harmful mutations of important cell proteins would lead to cessation of cell functions, but in case of proto-oncogenes it is the opposite, which means that their mutations lead to intensification of cell growth. Why they do this? If there were a cancer-making system inside a normal metazoan cell, the answer to this question would sound simply like this: those mutated proteins turn that cancer-making system on. However, it is not in the interest of an organism to have such a system of self-destruction. It is also very unlikely that an altered cell protein be capable of changing the whole chemistry of a cell to the extent to make it cancerous. One way of finding the connection between the proto-oncogenes and the initiation of cancer, in my view, is to look into the reasons behind the phenomenan of attachment of dividing cells to the extra-cellular matrix, the so-called social behaviour of the cells. I believe that when a cell within a tissue starts to divide, first of all it seeks the approval of the neighbouring cells to ensure that it divides within the context of the general design of the organism. It does so by using something which looks like an identification card. In this 'identification card' is the organism's design code encrypted. Once its identification card is being approved by the neighbouring cells, that particular cell loosens its contacts again and starts dividing. In any cell of a multicellular organism when a protein that is part of its design code is altered, that cell fails to recognize the design code of the neighbouring cells, and at the same time the alteration of such a protein makes it grow like an urprimordial cell, and invades the neighboring tissues; this is the way how a cell with a defective design code can initiate cancer.

To understand the uniqueness of a cancer cell, a distinction must be made between cells that can be reprimordialized and those that cannot. For the purpose of reprimordialization all design code proteins of the cell have to stay intact. A cancer cell, on the other hand, cannot be reprimordialized. The reason for this is because of the alteration of one or more proteins of its design code. Thus, I define a cancer cell as 'a cell with a defective design code'. Since the surrounding tissue cannot direct and control the growth of a cancer cell within the normal boundaries of the organism's structures, it grows to a cell mass that eventually invades the surrounding tissues and organs. Below I give a brief comparison between reprimordialization, deformity/non-viable body, and aberration (cancer).

Reprimordialization: Reprogramming of an existing design code into a new design code inside a differentiated cell.

Deformation/non-viable body: Alteration of the sequence of design code protein(s) inside an egg cell.

Aberration (cancer): Alteration of the sequence of design code protein(s) inside a somatic cell.

DESIGN CODE AND THE POTENTIAL FOR ADAPTATION

The potential for adaptation is common among all living organisms. This ability is usually recognizable in the flexibilities and variations the living organisms show with regard to their behavior, anatomy, and physiological functions in response to changing environmental conditions. At the molecular level, adaptability means autonomous rearrangement and reorganization of the organisms' macromolecules in response to the environmental factors that affect them. This involves, for example, modification and repair of DNA molecules, sequence and conformational changes of protein molecules, redesigning of protein molecules, reshuffling of cytoskeleton elements, and readjustment of metabolism. As it has been mentioned before, one form of DNA modification under cell control are DNA rearrangements using mobile DNA elements. This includes single nucleotide substitutions, insertions or deletions of large and small nucleotide sequences, and translocations of DNA sequences. In fact, living organisms have not only the ability to alter their DNA sequences, but also to produce new ones, even at a very fast rate, as it has been described earlier in connection with the production of new DNA blueprints for the variable regions of immunoglobulin molecules in vertebrates.

It is important to keep in mind that the changes the organisms introduce to their DNA sequence do not pertain to the sequences of their invariant proteins that are involved in encoding their designs. Therefore, I define adaptability as a potential in living organisms that enables them to use a variety of measures to survive and at the same time maintain their identities as distinct forms in an ever-changing environment. In each organism the potential for adaptation is determined by its design code. If the environmental conditions become so harsh that the organism fails to mobilize those measures, then it will die. While according to evolution theory adaptation would ultimately lead to evolution of one form into another form, at the cost of identity of the primary form, under the design code theory, however, transformation of one form into another is not possible; there are no drives in biological forms to move passively toward their own extinction.

Adaptability is not dictated by the environment

My notion of adaptability differs from the concept of adaptability of Darwin. He supposed that it works as changes that occur at random under the control of the environment, without direction. His point of view led him to the conclusion that, as he put it, "if they [those changes] were allowed to accumulate for many generations they lead to evolution of new forms." According to my design code theory, adaptability is not dictated by the environment; on the contrary, it is the potentials an organism uses to absorb environmental effects. They are manifestation of the organisms' own capabilities, whether it is a change in color or variations in size, shape, etc. Darwin's concept of adaptability is based on his interpretation of the works of breeders. He thought that "they unintentionally expose the organic beings to new and changing conditions of life, by which variability ensues, and among those variations they select those they want through breeding." I do not agree with his notion, because when we look at all dog breeds (one of the examples of domesticated products Darwin was referring to), they all are still dogs. The differences they show in shape or size are being brought about by the existing potentials within that biological form that is dog. To give another example, no matter how many varieties of wheat have been produced by breeders since the dawn of human civilization, they all are still wheat. When a breeder manages to produce, for example, a wheat variety that simultaneously encompasses three different qualities such as high content of protein, resistance to disease, and a relatively shorter maturating time, it is because of crossing between wheat varieties that maintain the potential for one or the other of those traits active at the time of breeding.

In nature, a famous example of adaptation is the case of the salt-and-pepper and black varieties of *Biston betularia* moth.

In pre-industrialized England this moth had a mottled-gray upper wing speckled with areas of black. The color and design of their wings provided an effective camouflage against the lichen-covered trees of their living environment. For the first time in 1840 moths with black variation of their wings were observed there. Half a century later, 98 percent of the moth population in the same area were with black wings. According to the design code theory nothing unexpected happened to those moths, because their design code enables them to react in that way. They obtained the black pigmentation because they had the potential for black pigmentation, i.e. its biochemistry, even in the originally white areas of their wings. Under conditions where the industrial pollution has increasingly darkening the bark of trees, on which these moths rest, the use of this potential allows them to adapt to this new environmental change, in order to reduce the chances of being seen by their predators. The same argument can be applied to answer the question why, for example, leaf-eating insects are green, or why animals change their coloration in response to the seasonal or instant changes in the intensity of light impact on the surface of their bodies.

DESIGN CODES DETERMINE THE PATTERNS OF REPRODUCTION

According to the design code theory the potential for sexual and asexual reproduction was rested in urprimordial cells, and for each form the method of reproduction is determined by its design code, as the spectrum of variation among the organisms indicates: sea anemones and marine worms can split into two half-organisms, each of them then regenerate the missing half. A solitary multicellular Hydra can produce offspring by budding. In sexual reproduction the separation of sexes into male and female is delegated by the design codes to the subordinate regulative systems. In mammals the protein which is required for determination of sex is a protein named *SRY*. If this protein is available during embryonic development, it acts on the developing gonads and leads to their differentiation into testes rather than ovaries. It triggers the differentiation of sertoli cells from one of the somatic cell lineages in the indifferent gonads (genital ridge).

Once sertoli cells begin to differentiate they must then signal to the other cell lineages that they should follow the male pathway, leading to the formation of functional testes. In the absence of this protein the cells of genital ridge differentiate into folicle (granulosa) cells. In *Schizasaccharomyces pompe*, its cells in addition to transitions between asexual and sexual behavior, also show a transition between the opposite sexual statutes. In this organism a shift between two sexually opposite cells has been found to be controlled by the cell type-specific proteins. A cell in order to change its type alters the sequence of its type-specific protein through conversion of its sequence. In marine worm, Bonellia, the future of an egg whether to become a male or female depends from whether it falls on the proboscis of female or not. If it does, then it develops into a male worm, otherwise, into a female one. In some alligators it has been shown that at 30°C eggs develop into females; at a temperature above 34°C they develop into males. There are lizards which consist only of female individuals and reproduce without mating. Egg cells can, for example, develop without merging with sperm, a method referred to as parthenogenesis. In rotifiers, nematodes, mites, and insects haplodiploidy seems to determine the sexes, in which haploid males develop from unfertilized eggs and diploid females develop from fertilized eggs.

In this place, referring to the known details of reproduction of *Caenorhabditis elegans* I will explain that the process is in all its details designed by this organism's design code and nothing else, such as accumulation of tiny changes as the supporters of evolution theory might assume.

An adult *C. elegans* is about 1mm long and consists of only some 1000 somatic cells and 1000-2000 germ cells. This animal is transparent and its cells can be watched as they divide, migrate, and differentiate. The cells of the germ line separate from the somatic lines at an early stage of development. It has two sexes: a hermaphrodite and a male. The hermaphrodite sex can be viewed as a female that produces a limited number of sperm. She can reproduce either by self-fertilization using her own sperms, or by mating with a male. The egg-laying orifice, the vulva, is a ventral opening in the hypodermis (skin) formed by cells that arise from this layer. A single non-dividing cell in the gonad, called anchor cell, attaches or 'anchors' the developing vulva to the overlying gonad to create a passageway through which the eggs can pass to the outside world. Laser destruction studies have shown that the anchor cell is responsible for inducing the three nearest hypodermal cells to the gonad to form a vulva. The inducing signal from the anchor cell ensures that the vulva develops in exactly the right place in relation to the gonad. If the anchor cell is killed, these cells, instead of following a vulva-lineage, give rise to ordinary hypodermal cells. If all of the gonadal cells except the anchor cell are killed, the vulva still develops normally, indicating that only the anchor cell is necessary for this induction. This information when viewed in the context of design code theory means that the function and positioning of anchor cell between the gonad and the hypodermal cells is only possible when it is already programmed in the framework of the general design of the form. For the support of this notion I give the following analyses. Supposedly, if it were a mutation that turned one of the cells into an anchor cell, then, it must have been followed by a second mutation that enabled the anchor cell to produce a new protein intended to create a passage (vulva) in the hypodermis layer. In order for this structure to be formed, some other mutations within the participating hypodermis cells must have been occurred, too, making them, among others, capable of producing membrane receptors to bind the anchor cell's protein. This is still the beginning. Certainly, many more random mutations and so-called natural selection work would have been needed to make the hypodermal cells understand the purpose of those changes and initiate the necessary intra-, and intercellular responses toward formation of the vulva. This is very unlikely that such a coincidence ever happened. Now, let us assume, for the sake of clarification that at some stage during the development one of the embryonic cells was mutated into a germ-line cell to produce gametes, but there was no anchor cell and no orifice. In a situation like this, the animal was actually wasting its energy resources producing gametes that could not get into the outside world. This would not be a sign of fitness for an organism to waste its energy by producing gametes in such an ineffective way. The fact that all morphological and functional aspects of sexual reproduction can be expressed in a coordinated manner is because the design code of the organisms program them like this.

DESIGN CODES PRESET ORGANISM-SPECIFIC PATTERN OF BEHAVIOUR

For their survival living organisms need resources of nutrient supply. Those resources are subject to extreme variations. Thus they have to recognize every possible variation of food in their surroundings and know how to deal with them. In case of single-celled organisms all means and strategies of survival are in one cell. In metazoa the means of interaction with the environment have been so designed to be distributed over groups of cells and organs. The reasons for this are obvious. In vertebrates and invertebrates, due to differentiation of their tissues, most of their body parts have no direct access to the outside world. Therefore, information coming from the environment must be taken by cells, tissues, and organs that are exposed to it and relayed to the rest of the body. At the same time information must be transmitted in the opposite direction, that is from inside to the outside. To this comes that these organisms have to analyze the incoming and outgoing information back and forth. In this sense, nervous systems [in organisms with more complex designs] are restructured versions of the urprimordial cells' communication and decision-making system. When these faculties obtain their ultimate shape at birth, or even after birth, given that no outside help is involved, we tend to call them instinct or unlearned behavior; when these faculties take shape after birth through learning we call them learned behavior.

Behavior like any other functions serves a purpose, and its components must be put together right from the start. For example, let us consider a bird that has to feed its chicks. In order to be able to carry this job effectively, first of all consistency between its abilities of flying, finding food, building nest, and recognizing the cries of its chicks demanding for food must be established. The bird also needs to know that its chicks must be fed on a regular basis for a certain period of time. These are possible thanks to their design codes. They are not acquired through accumulation of tiny changes, as the supporters of evolution theory assume. The coherence in sequence and purpose which we see in instincts contradicts such an assumption. Referring to the relation between larva and 'adult' fly, which I have described earlier, the larva has its own specific behavior and the 'adult' fly its own specific ones. At the moment a larva disintegrates, its instincts disappear with it. When through reprimordialization an 'adult' organism emerges, it is furnished with a completely new set of instincts.

An excellent example of instinctive behavior is that of the

solitary wasp. This wasp before laying an egg, the first thing it does is the making of a borrow in the ground; then, in order to secure the right source of energy for its offspring's development catches a caterpillar and places it in the burrow. After laying a single egg on that caterpillar the wasp closes the mouth of the burrow and leaves the egg unattended. The new wasp that emerges from the burrow, without ever knowing its parent or having watched any other wasp, manages to fly and find nutrients. When the time comes to lay its own eggs, it proceeds exactly the way the mother did, providing the eggs with sustaining nutrient sources, when possible, with caterpillars. She obviously knows that a caterpillar is a good source of food and will suffice the needs of its egg during its development. The wasp knows about the action of its venom when stinging the caterpillars, and then how to place the egg on it. The wasp knows how to prepare the proper cement for closing the burrow. This complex and coherent pattern of behavior which foresightedly guarantees the survival of this wasp must have been designed and put together in one step. If one extrapolates this pattern of behavior to human dimension, it would require years of training for an adult human individual to learn abilities equivalent to this. Another example of extinctive behavior in which foresight is involved is the reproductive behavior of tree frogs Chiromantis. They lay their eggs on tree branches above water. As the eggs are produced, the mating frogs use their feet to beat the eggs and seminal fluids into froth. In this protective environment the eggs fertilize and develop into larvae. When the foam drops down from tree branches, they carry with them the larvae into the water, where they get a different design code, thus turn into frogs. An strikingly complex form of instinctive behavior with high degree of foresight is that of caterpillars that develop from eggs of butterflies Nymphalis. First, they glue themselves to a leaf or twig by the cremaster (a spiny process at the end of their bodies) hanging head downward. If it were not for their foresight, at the time the outer shell of the caterpillar comes off the emerging pupa would fall to the ground. Surprisingly, this does not happen, because the cremaster with which the caterpillar glued itself to the twig or leaf does not come off with the shell, so it keeps the pupa still fixed in the same position. The pupa subsequently reprimordialize into a butterfly, which emerges from the capsule and flies off. This time the capsule comes off together with the cremaster, otherwise the butterfly could not fly away.

After getting some insight into the nature of some instinctive behavior, I want to point to an apparent connection between instinct and the level of competence for survival soon after birth. It seems that the more a living organism is equipped with instincts, the more capable it is to stand on its own feet. For instance, if we compare the number of instincts of a newborn human baby with the number of instincts of a newly hatched solitary wasp and look at the number of their skills at that point of their lives, we will find a clear correlation between the two parameters. The only instinctive behavior that newborn human babies show at birth is sucking; so with respect to skills they are very helpless. They are even unable to search for food on their own for years after. On the other hand, as mentioned above, a newly hatched solitary wasp thanks to the higher number of instincts that it has, posses so many skills that are enough to serve all its needs during its life.

Man and those animals that are equipped with only a small number of instincts have to learn the techniques of survival from those animals that already have such techniques; either through direct mimicking, or by means of learning from the experiences of the members of their own populations who learned those techniques before them. On the basis of evidences we have today, numerous patterns of behavior in an adult human being, no matter how complex they look, are already identifiable among many insects and animals. Some of the techniques we are using have been utilized by them for million years. For example, the beginning of our skills to construct bridges may go back to a few thousand years, but ants however can build even living bridges. Our using of high frequency sound waves to identify objects hidden from our eyes may be a century old, but bats, for example, are so designed to produce and use such waves. Some moths upon which bats feed are even in possession of a sonar jamming system. This mechanism helps the moths to escape bats when they close in on them guided by their sonar system.

Biodiversity has its roots in design codes

I n the previous chapters, design code theory delivered insight into the basics of a broad range of biological issues. In this chapter, with the help of my theory I discuss the issue of biodiversity. I hope the reader appreciates the solid arguments I offer rather than seeking help from imaginary forces and actions such as 'random selections', 'accumulations of tiny changes' as the supporters of evolution theory have been used to resort to.

According to the supporters of evolution theory, due to the ever-changing environment those organisms that are better adapted have a higher chance to survive and reproduce successfully, thus pass their superior qualities on to their offspring, and as a consequence, each generation will be slightly different from the previous one. They believe further that the useful variations are preserved while those harmful are destroyed. They think that the useful variations occur by two kinds of forces, one which causes variations and exists outside the living organisms and the other, the so-called natural selection, determines which variations are favorable and therefore should be preserved. For instance, they conclude that it will be possible that a flower and a bee slowly become, either simultaneously, or one after the other, modified and adapted to each other by the continuous preservation of all the individuals that presented slight deviations of structure mutually favorable to each other. In their view, environment is acting in two ways: first, directly on the whole organism or on certain parts of it through increased use and disuse of those parts; and second, indirectly by affecting its reproductive system. With regard to disuse of parts they argue that this would lead to their reduced size. They believe, for example, that this happened to the wings of penguins and ostrich. The former because of living on the ocean islands have seldom been forced by beast of pray to take flight, ultimately lost the power of flying; the latter because of the increased size and weight of its body became incapable to fly. Comparing a lizard with a snake, according to the evolution theory, due to disuse the snake lost its limbs. Concerning the effect of increased use of organs they refer, for instance, to the size and shape of giraffe's neck and tail, and the long beak of hamming birds concluding that the first ones occurred as a result of sustained effort for better reaching the canopies for leaves and to drive away the flies, respectively, and the second one resulted from a tendency among the ancestors of this bird to adapt their beaks to the form of the flowers from which they collected nectar. With respect to the indirect changes of reproductive system their notion is that this kind of variability is inducible because of the extreme sensitivity of the reproductive system to environmental changes. By the following example they tries to offer an explanation for what they mean under 'the extreme sensitivity of the reproductive system to environmental changes': regarding the size of jaws and teeth of man, the free use of arms and hands have led in an indirect manner to modifications of those structures. The male forefathers of man were furnished with great canine teeth, but as they gradually acquired the habit of using stones, clubs, or other weapons they used their jaws and teeth less and less until their jaws together with their teeth became reduced in size.

Those arguments mentioned above cannot hold, because as we know there are no ways that changes acquired by a developed individual through disuse or increased use can affect germ cells; therefore, it is impossible for acquired changes to be transferred to an organism's offspring. Looking at the morphology and behavior of giraffe, humming birds, penguins, ostrich and man through my theory, we would see other reasons behind those body designs. Giraffe has a long neck, because its design code determines this way. Therefore, it must be understandable that this animal with its long neck favors the leaves at the top of trees than the low vegetations on the ground, otherwise, because of the unique design of the animal, the latter alternative would be strenuous for it. Concerning the humming bird, let us assume for a moment that the imaginative ancestors of these birds were originally seed-, or fruit feeders that due to a long-lasting drought period, according to the evolution theory, were forced to migrate out of their habitat;

then, after exhausting search for food they ultimately found a place where plenty of nectar producing plants were available. What happened next? They would have died if stopped continuing their search, because there is no mechanism available in nature that could have changed the entire anatomy and physiology of those birds to make humming birds out of them. First, in order to reach the inside of the flowers, they had to develop long beaks. Second, they had to maintain a steady position in the air, for this they had to reduce the size of their bodies and at the same time increase the stroke frequency of their wings. According to my theory, humming birds have no ancestors looking different from them; they were emerged with a body design perfectly suitable for collecting nectar. The design included all the necessary morphological, physiological and biochemical prerequisites characteristic for a humming bird. Regarding the wings of ostrich and penguins, we have to realize that wings are needed not only to help birds fly above the ground, but also to serve birds like ostrich to get more upwind in order to run faster. Wings can also be designed in such a way to be fit for swimming, as it is the case with penguins. One should bear in mind that it is not only the presence of wings and feathers that characterize these birds, but also the wholeness of their forms including their physiology, morphology and behavior that make them penguin. To get to humans, there is no way by which the jaw and teeth of man could be evolved, directly or indirectly, from those of an ape. It is also unthinkable that man had forefathers that were furnished with great canine teeth, and they gradually lost the original size of their teeth together with the size of their jaws.

It is impossible that changing environment could change the anatomy of a living form. Behavior and structure in all metazoa are programmed together by their design code.

Similarity of sequence between homologous proteins and estimates of the rate of DNA mutations provide no evidence of phylogenetic relationships

Recently protein sequence analyses have been used to construct phylogenetic relations among living organisms. It has been assumed that the stronger the similarity between the sequences of two homologous DNA and/or proteins, the nearer is their relation; conversely, the bigger the difference between the two sequences, the further apart they are from each other. Beside this, in some other studies attempts have been made to figure out the rate of DNA mutations in living organisms. Assuming that DNA mutations occur at a regular rate, it has been expected to find a proportionality between the time elapsed after branching of related organisms from their last common ancestor and the number of mutations fixed in their DNA by the so-called natural selection. In the following, I will bring arguments to show that these assumptions are not correct.

1. The differences found in the sequences of homologous proteins in living organism do not reflect the differences they show with regard to their morphology, physiology and behavior. For example, the difference between the amino acid

sequences of corresponding proteins from apes and man seems to range, according to the type of protein, from zero to less than one percent, but despite this small difference, they do not look alike or behave alike. In another example, on the basis of similarities found between cytochrome c sequences of cow, pig and sheep they should be near relatives; as well as camel and whale, for that matter. Furthermore, as with haemoglobin it has been shown, the function of various mammalian haemoglobins does not vary significantly, despite differences between their sequences.

2. Living fossils (living organisms that inhabit the earth for a couple of hundred million years or more) are still the same as their early ancestors.

3. Regarding the estimates of the rate of amino acid substitutions I want to mention that changes in the amino acid composition of proteins in living organisms caused by environmental factors, as I mentioned before, have no relevance with regard to their diversification. Nonetheless, for the sake of completeness of arguments I discuss this matter below on a number of proteins that have been used for calculating such estimates.

a) A given amino acid site in a protein may have been changed more than once, but the outcome could be only one observed amino acid difference, due to the return to the original amino acid by the second mutation. Similarly, the organisms that are compared with each other may have incorporated similar changes at the same site, thus ending with no difference.

b) Sequence analyses of proteins do not reflect the entireness of mutations occurred in a protein. For instance, in case of haemoglobin it has been shown that changes that occur in the interior of the molecule are generally harmful, and most changes occurring in residues on the exterior of the molecule appear to be less harmful. This is so, because the substitutions that occur in exterior sites of this protein have no impact on its function, while substitutions in interior sites impair its function. In cytochrome c about 74 to 81 of its amino acid residues are variable, and their substitutions apparently do not affect the function of the protein. The rest are invariant and are important, because they are needed for combining with the heme group, or for interacting with cytochrome c oxidase. Histone-IV protein apparently shows only two substitutions in peas and cattle, because virtually all amino acids residues in histone-IV are invariant.

c) The matching property of homologous protein sequences may be thrown abruptly out of phase by deletions of one or more amino acids. For instance, human and carp haemoglobins with regard to residue 47 in the molecule differ from each other. At this site an alanine residue is present in the alpha-chain of the carp, but a gap in human.

d) Because of the degeneracy of the genetic code, there are 61 amino acid-specifying codons, instead of 20. Codons that specify the same amino acid are called synonymous codons. These codons differ from each other in their third nucleotides. Therefore mutation of one synonymous codon into another causes no amino acid substitution in the sequence of a protein.

From the information provided by sequence analyses one can draw these conclusions: 1. the existence of phylogenetic relations cannot be proven. 2. Proteins in living organisms, whether they have undergone sequence variations or not, seem to have one common origin, and that as I assume is the urprimordial cells. 3. Variations detected in protein sequences between different organisms do not account for their morphological differences. 4. No single form-specific protein has been found, because according to the design code theory it does not exist. As it has been described in chapter one, designing of forms is based upon proteins which are invariant and were already present in urprimordial cells.

'Form' versus 'species'

The facts described above underline the need that the classification of living organism has to be based on criteria that distinguish forms from each other. The present choices of criteria for classification of living organisms are subjective. The system of naming and classifying living organisms that is still in use today is based on the degree of outer resemblance, starting from the more basic category named species, toward exceedingly larger categories: the species into genera, genera into families and these into orders, orders into classes, and the latter into kingdoms. Species is defined as distinguishable groups of genotypes that remain distinct in the face of potential or actual hybridization and gene flow. It has been turned out, however, that hybrid speciation, even without polyploidy, is more common in plants and animals than were assumed. Now, there are about a dozen different definition of the concept of species, pointing to the vagueness of this term.

Not only are the criteria used for assigning organisms into species or subspecies categories are vague, but also those used for placing the organisms into higher taxons. Based on their anatomy nematodes were placed close to the 'root of the tree of the animals'. They lack a body cavity called a coelom, which is found in molluscs, insects and vertebrates. Using molecular data they were put together with insects to a group called Ecdysozoa [its members grow by moulting their outer layers]. To justify this shift, it has been assumed that nematodes apparently lost their coelom.

As we see the criteria by which the taxonomists judge the degree of resemblance between organisms depend from the taxonomists' perception about their importance. If taxonomists find a simple character to be common among a great number of forms, they use it as one of high value; on the other hand, when an outstanding feature happens to be less common, they use it as one of subordinate value. Furthermore, by assuming phylogenetic relations among living organisms it could happen that a simple trait be considered mistakenly as evidence of relationship between some living organisms. To name a few: the manner in which the wings of insects are folded, the mere difference in color in certain algae, the nature of the dermal covering (hair or feathers) in vertebrates, and tiny middle ear bones in mammals etc.

Using differences between homologous sequences of DNA as an instrument of classification of living organisms is equally inadequate. So far some 1000 different organisms' genomes have been sequenced. Though the data obtained are worth spending time and effort, but lacks the least consistency with regard to establishing relationships. As one can see, neither anatomy nor sequence data establish something which does not exist, namely phylogenetic relationships. Changes in sequences of DNA, in whatever large number it might occur, have not been found to be related to the designs of the organisms. Further, the existing variability of satellite DNA sequences among living organisms led to the belief that this fraction of DNA might be related to 'speciation'. These variations, however, are non-specific even to the extent that members of the same living organism vary with regard to their satellite DNA. The structural and numerical chromosome changes also cannot explain the differences between living forms. First, they are very irregular and second, if there were any connection between them and 'speciation', one would expect those connections to be identified. In experiments where the effects of change in the number of chromosomes were studied in salamanders, it was found that animals with haploid (11 chromosomes) and pentaploid (55 chromosomes) sets of chromosomes were of the same size and appearance; they also did not differ with regard to the morphology of their internal organs. While the cells of the pentaploid salamanders were roughly five times as big as those of the haploid ones, the number of cells in them were one-fifth of the latter. Therefore, reduced cell numbers are compensated for increased cell size. This observation is consistent with my theory of design code that indicates that neither alterations of the sequences of DNA, nor changes in the number and size of chromosomes can lead to a change of one biological form into another.

From the above deliberation I conclude that the classification of living organisms must be based on criteria that can distinguish the forms definitely from each other according to their design codes. After this has been established, then different populations of any form should be catagorized according to traits such as size, color, etc.

Fossil finds from the perspective of design codes

In this part I review some of the vertebrates' fossil data and their interpretations by palaeontologists, whereby I will try to differentiate between assumptions that are based on facts and those which are pure speculations. Fossils when well preserved and their dating is reliable can give us valuable information about the structures and designs of the living organisms that inhabited the earth in the past. Fossil finds are also important in telling us about the way of life of those organisms and the events that affected their lives. One has to keep in mind that the records of fossil finds are not continuous, and they are characterized by gaps, which are generally very large (ranging between some million of years to over hundred million of years). For this reason, one should hold back from using fossils to 'reconstruct phylogenetic relations' between living organisms.

Vertebrates' fossils constitute a major focus of interest for palaeontologists. It has been thought that the remote ancestors of all vertebrates including man are the chordates, animals with a hollow nerve cord, gill slits, and notochord. The oldest fossils of chordates dated back to 600 million years ago. Those fossils are very similar to the living lancelets known as Amphioxus. Chordates on their way to vertebrates, as it has been assumed by paleontologists, had given rise to the early craniates like hagfishes. From those animals, they argue, fifty million years later supposedly the first vertebrates have been evolved, characterized by having segmented cartilages protecting their spinal cords, with lampreys as their living representatives. Paleontologists imagine further that those early vertebrates then gave rise to vertebrates with jaws and paired fins. The earliest representatives of the latter dated back to 400 million years ago, with sharks as their living representatives. Those animals in their turn are believed to be the ancestors of the bony fish, the first vertebrates with internal skeleton and lungs or swim bladder. It has been further assumed that subsequent diversification of bony fish led to evolution of ray-finned and lobe-finned fishes. While further diversification of the first ones supposedly led to modern fish varieties, the diversification of lobe-finned fishes led to amphibians, the first tetrapods on earth, 370–360 million years ago. The assumptions reviewed above are very speculative. The following analyses show, why.

1. If any of the new structures that replaced already existing structures in any of those ancestors was making them more suitable for survival, why should animals with the older structures still exist. With or without cranium, with or without jaw, with or without inner skeleton, with or without lungs, their representatives are still living in waters all over the world.

2. The speculative nature of these assumptions is also reflected in the inconsistency about the age of the fossils. The oldest fossil of hagfish found dated back to 350 million years ago, while those of the first vertebrates and jaw fish are 500 and 400 million years old, respectively.

3. How did a bony fish acquire lungs, and why this had to be selected? Some palaeontologists tend to believe that there must have been a violent alteration of seasons leading to scarcity of fresh water, so that the bony fishes found themselves in a stagnant pool and suffering from oxygen shortage, having no other choice but to develop lungs. But sustained drought is impossible to be the cause for lungs to develop. The most likely consequence of a sustained drought would have been extinction of those fishes.

4. With regard to the limbs of the first vertebrates one has to

ask this hypothetical question. Did limbs evolve while those fishes were still living in water, or later when they moved on land? If it happened in water, there was no benefit in this for them; if this change took place on land, then why they had to leave the waters in the first place. In addition, limbs cannot evolve because of any excessive use of fins for walking; developmentally it is impossible. According to the design code theory, however, fins and limbs each belong to a different form, which are designed and emerged independently.

Fifty million years later, supposedly, a small creature named stem reptile evolved from among amphibians to give rise to reptiles known from Mesozoic era (covering 150 million years) and the ones that appeared later. Those were reptiles of different sizes and appearances such as dinosaurs, the various types of marine animals like marine turtles and sea snakes, crocodiles, flying reptiles, etc. Here again, the suggestion of an existing phylogenetic relation between amphibians and those creatures is unfounded for a number of reasons.

1. Common sense does not accept this assumption, because why animals that had both water and land as living environment at their disposal should abandon that double opportunity and restrict themselves to land only and become reptiles.

2. The assumption that some of them retreated for the second time to water (to build, supposedly, the marine forms), to the same environment that was not good enough for them

at the beginning, adds to the unlikelyness of the idea that those amphibians left water in the first place. Therefore, the assumption of a phylogenetic relation between amphibians and reptiles is a very speculative notion.

In the absence of a common ancestor for reptiles, as a matter of fact, the existence of phylogenetic relations among reptiles themselves is also unrealistic. Of course, they share a number of characteristics like the structure of their jaws, vertebrae, etc., but the existence of similarities between parts of their anatomy in my understanding is not justified to draw the conclusion that reptiles are phylogenetically related, i.e. the later forms were evolved from the earlier ones.

From the fossil finds it has been assumed further that the earliest ancestors of birds and mammals were lizard-like creatures that first appeared in Carboniferous period 310 million years ago. They then supposedly divided into Diapsid (sharing two temporal fenestrae) and Synapsid (sharing the lower temporal fenestra only) reptiles, whereby birds became part of a Diapsid clad, and the mammals part of a Synapsid clad. According to a different guesswork birds and mammals, because of sharing the homeothermic characteristics, were considered as sister-groups diverged from a common ancestor about 225 million years ago. Both assumptions have one thing in common: they are only guesses, as the following arguments will show it.

A bird by no means can evolve from a reptile, no matter

what anatomical similarities they might have in common, because the design of a bird requires a totally different program than that of a reptile. In order for any reptile to become a bird, that organism must have acquired all anatomical, physiological and behavioral aspects that are characteristic for a bird, simultaneously. This is something absolutely impossible to have taken place. The major characteristics of birds' skeleton are the hollow bones, three functional toes, half-moon-shaped wrist bone, fused clavicles, keeled sternum and short tail. There are also suggestions that the size of avian genomes, smaller compared to other amniotes, had helped birds to fly by reducing the metabolic costs associated with having large genome and cell size. However, small genome size was also common among some reptiles before emerging of birds. This adds one more to the birds' attributes such as feathers, claws, pulmonary characteristics, and parental care and nesting to be possessed by some reptiles before them. However, the existence of dinosaurs with claws, feathers and other birds' attributes neither made them birds nor ancestors of birds; they are designed in that way. Thus, the only way for birds to emerge must have been through an independent design coding that gave them all those major characteristics simultaneously as essential components of their design.

Now I want to analyze the situation about mammals. First, it needs to be pointed out that mammals is a name given to a large group of diverse animals to indicate their having milk-producing mammary glands. Since mammary glands developmentally turn into functional state years after birth, they should not be used as a criterion of relatedness of all those animals. Second, the idea that mammals because of the similarities of their detached tiny middle ear bones are related is also weak. Third, there has been not a single fossil which could show a link between any of the various forms that are included in this group. The living monotremes represented by the duckbilled platypus and the spiny anteater or echidna, are egg-laying animals. The marsupials are a diverse group of animals that have been called so because of their females having a pouch in which they carry their immature foetuses. The placentals consist of organisms that are in many respects different from each other. They are grouped into eighteen orders which include insectivores, edentates, rodents, logomorphs, bats, primates, artiodactyls, perisodactyls, carnivores, elephants, cetaceans, etc. The assumption that all these animals have as their predecessor an insectivore (animals not very much different in size from Tupaia), which apparently was living in abundance during Cretaceous period is unfounded.

Thus, there is no evidence which could indicate that these orders have any phylogenetic relation with each other, but also no proof that the animals in each of those orders are phylogenetic related to each other. For instance, in all those various forms of animals that are classified under artiodactyls (even-toed hoofed mammals) the resemblance of their limbs simply means that the design of such limbs fits the design of the animals to which they belong, nothing more.

The phylogenetic reconstruction attempts also seem pointless, when one looks at the primate fossils. It has been assumed that the organisms included in the order of primates such as monkeys, lemurs, man, apes, etc. are the descendants of an extinct prosimian, which was presumably living during Paleocene in Africa 70 million years ago. Oddly enough, such assumptions are being made despite the fact that in Africa following *Eocene* no fossil record of prosimians was found. Furthermore, there exist no fossil which could link the prosimians themselves to insectivores; at least a gap of twenty million years exists between the recorded fossils of insectivores and those of the early prosimian.

Design code theory and the emergence of man

Regarding the emergence of man it has been assumed by the supporters of the evolution theory that man is evolved from ape-like creatures, supposedly in Africa during Pliocene and Pleistocene, despite lack of any fossil record of apes in Africa during Pliocene and Pleistocene. To compensate the lack of evidence, they imagine that at least one African ape left the continent about 18 to 20 million years ago, but then after its descendants eventually gave rise to an array of early apes in Asia and Europe one of those Eurasian apes moved back to Africa by about 10 million years later and became the ancestor to the living African apes. Later due to some hypothetical events some of those apes supposedly came down from the top of trees, whereby in the new environment their struggle for survival required increased use of their front limbs. At the end, this supposedly led to a new creature that could walk on two limbs, referred to as australopithecine. In the Following I present some arguments against such stories.

1. Even if it were possible to prove that some 3-5 million years ago there was a climatic change, or any other event such as a volcanic eruption in Eastern Africa, leading to shrinkage of forests, it is unlikely that apes instead of retreating into the still remaining forests dispersed themselves into the open plains. If this were what the apes did, most likely they could not survive very long.

2. By comparing a very large number of matching genes from human and chimpanzee it has been indicated that favorable mutations in chimpanzee genes were higher in number than similar mutations in human genes. This kind of finding in the context of the design code theory means that those alterations have been taking place under the control of the organisms and not the environment. Otherwise, if human were supposed to be evolved from chimpanzee, then there should be no favorable mutations for the latter and this organism should stay where it was. What would be the purpose of those favorable mutations, if one assumes that they represent the 'selected' ones by 'natural selection' from among the 'randomly occurring tiny changes'? Is 'natural selection' on the path of planning to transform for the second time chimpanzee into a different organism, better than human? However, by seeing the inner potentials for flexibility in living organisms as the factor behind the favorable mutations, it is conceivable that those are offering the chimpanzee the chances to become stronger in maintaining its own form identity.

3. The oldest bone remains which some paleonthologists consider as evidence of an upright walking ape is the three-million-years-old bone remains of an ape-like creature known as Lucy. It is the skeleton of a young adult female creature with over three feet tall; the skull is extremely fragmentary and has an ape size. The youngest bones belonging to a similar creature is about one million years old. In all those cases perhaps the shape of their teeth or the curve of their jaws show some similarities with that of man, but those similarities are not unusual according to my design code theory. Therefore, those skeleton similarities offer no ground to call them the remains of an ape-like human ancestor and as evidence of link between human and apes.

Just to show how speculative the stories about australopithecines are, in this place I consider it appropriate to mention a story about the interpretation of some bone finds ascribed to those so-called human ancestors. At the excavation site of australopithecine at Swartkan from the bone remains of some other animals piled up beside those of australopithecine one paleontologist concluded that they were resulted from the activities of bloodthirsty australopithecines; another one looking at the same bones later found them as leftovers from the meals of carnivores. In addition, since the bones found in that deposit were blackened, this was first thought to be due to the use of fire, later the deposition of manganese on the bones was found to be the cause of the blackening. The only assumption which might possibly be correct about australopithecines is that such creatures were living 4 to 1 million years ago on earth, and then for whatever reason they became extinct. The presence of sharp flints at the sites of australopithecines, even if it might indicate their familiarity with tools, has no relevance as proof of their relationship with neither man nor apes.

4. If the human skull and brain were evolved from the skull and brain of an ape-like creature, it needed to be explained how it occurred developmentally, and how the interconnection between the development of the two structures were established. Unfortunately, questions of great significance like this have been left almost untouched. In the following, I will explain that the skull and brain of man are not evolved from an ape-like creature.

In a study it has been argued that a reduction in sphenoid bone size may account for a rapid evolution of modern human cranial shapes. This kind of notion does not recognize the fact that developmentally the size of the skull cannot be determined by random changes in the size of the sphenoid, or any other bone, since factors that determine the size of the head and the brain seem to be developmentally interconnected. Let us ask this hypothetical question: was it in case of those ape-like creatures their attempts to walk on two limbs, or their attempts to use their hands that led to the enlargement of their brains? None of them is the answer, because there are no ways by which those attempts could influence the generative cells of those apes to lead to the development of individuals with larger skulls and brains. Studies on vertebrates' embryos indicate that cells from the dorsal ectoderm give rise to neural tissue. As it has been shown in experiments with zebra fish, in order for the dorsal ectoderm cells to develop to brain tissue the cooperation of a small group of cells located in the prospective head region of the gastrula is required. Without these cells the formation of brain tissue is disturbed. This study also indicates that in a developing embryo the factor which determines how large the brain should be depends from the size of the head region. It seems that not only the formation of the brain but also its size is jointly controlled by the cells of the head region. Having this correlation in mind, there is no possibility for any attempt of upright walking, or free use of hands to influence an already established relation of body parts. This cannot be achieved through accumulation of tiny changes. Thus, it is impossible for human to be evolved from apes or ape-like creatures.

In my opinion, human is designed independently through reprimordialization, thus have its own specific design code. Their history starts with Neanderthals. They were living for about 200000 years and we are their offspring, no matter how tall and how small we look or what the color of our hair and eyes are. Based on skeleton finds, their skull bones might have shown slight differences in size and shape when compared to recent generations of humans, but their form is recognizable as a human form, with a brain size of about 1400 cubic centimeters. In comparison, the size of the brain of those ape-like creatures that according to different estimates were living between 4 to 1 million years ago was approximately 400 cubic centimetres. If Neanderthals were the result of a further evolution from australopithecines, separated from each other by about one million years, then how could the 1000 cubic centimetres jump in the size of the Neanderthals' brain be explained? Since australopithecines with their 400 cubic centimeters brains were supposedly walking upright and using their hands, then one may conclude that such a rapidly brain enlargement of 1000 cubic centimetres in Neanderthals must have been exclusively required for making some artificial stone tools, something very unthinkable. Moreover, since Neanderthals were humans as we are and had the same brain size as we have, so we should be able to identify this 1000 cubic centimetres tool-making area in our own brain. Where is this area? In fact, such an area in our brain does not exist. One can continue arguing that if Neanderthals for making their tools needed 1000 cubic centimetres larger brains than the australopithecines, then where is the amount of brain enlargement needed for the development of language, agricultural and architectural performances by humans in the last few thousand years. There is, however, no evidence that any of the recent human performances coincides with any brain enlargement. Thus, it must be concluded that no form of activity could make a Neanderthal brain out of an australopithecine brain. Therefore, according to the design code theory, Neanderthals emerged independently and had no relation with the australopithecines.

From these analyses it becomes clear that the tendencies to interpret phylogenetic relations between fossil finds by putting emphasis on some selectively chosen anatomical features are groundless. If there were any such relations involved in diversification of living organisms, then at least concerning the more recent organisms one should be able to find first of all hundreds of examples of intermediate fossils between any two organisms considered to be related, and then the findings be consolidated on the basis of detailed developmental comparisons. It is very superficial to put existing living organisms with very different morphological, physiological and behavioural characteristics in one group and call them relatives because of sharing some anatomical or physiological features such as, for example, the mammary glands in case of mammals, or homoeothermic feature in case of birds and mammals. It is equally not correct to assume that because of the presence of their extra-embryonic membrane (the amnion), reptiles, birds, and mammals are relatives.

Summary of the Theory

1. Cell proteins are very intelligent molecules, having the ability to code the designs of living organisms, recognize events and memorize their sequences.

2. The emergence of a biological organism requires a design code, made by invariant proteins. This code whether it is made inside an urprimordial cell or any other cell determines the design of the form (organism) to be emerged from those cells.

3. An urprimordial cell is a cell that contains the basic functions of a living cell, such as growth, reproduction (sexually and asexually), signal transmission, communication with the environment, and the potential for cell-to-cell association.

4. A cell with a design code that leads to the development of a form is a primordial cell. It is a cell that in addition to the basic functions and potentials of an urprimordial cell also specifies the specific functions, behavior, and the layout of the structures of the organism to which it develops.

5. In addition to primordial cells any other cell (differentiated or embryonic) can also be turned into a primordial cell through reprogramming of an already existing design code, given that none of the invariant proteins that have to take part in the new design code is altered. I call the process reprimordialization.

6. In a differentiated cell or in an embryonic cell, when any of the invariant proteins that participate in their design code is altered the design of those cells destabilizes, by which they turn into cancer cells.

7. Transformation of one form into another is impossible. Form-specific proteins do not exist. Alterations of DNA sequences and variations in size and number of chromosomes are not the cause of diversity of biological forms.

8. Adaptability is a potential that allows an organism to undergo functional, behavioral, and morphological changes within the boundary of the form's design.

CONCLUSION

In this book I presented a comprehensive biological theory of design code that is able to answer a number of still unresolved questions in different areas of biology, such as the origin of forms' designs, biodiversity, development, initiation of cancer, and the origin of animal behavior. Some major aspects of this theory are the following.

I. We want to know what is it at the molecular level that makes the living organisms to be different from each other, such as, for example, a man to be different from an ape. This is understandably a very challenging question. What is the right answer?

Looking at the uniqueness of each biological form, I assume that in the absence of forms' specific proteins the design of a form must be determined by a design code that is made of invariant proteins, which are undergone no changes in critical parts of their sequences since the primordial times; they can combine with each other in thousands different ways (depending from the number and the sequence of their alignment). The mechanism in charge of combining these proteins in specific patterns to produce design codes I call primordialization. During this process urprimordial cells are turned into primordial cells, so that different primordial cells posses different design codes. These programs can also be encoded in cells that are equivalent to urprimordial cells. Interestingly, it has been reported recently that metazoans' proteins such as cell signaling and adhesion proteins can have their origin from coenoflagellates: Expression in these organisms of proteins involved in cell interactions in Metazoa indicates that these proteins already existed before the emergence of animals. When this is the case, then one has to go one step further and ask the question from which organisms the coenoflagellates obtained their proteins. Design code theory says: from urprimordial cells; certainly not so that those proteins were being passed from protozoa to metazoa, but so that protozoan and metazoan all receive them directly from urprimordial cells or cells equivalent to them at any time when a new distinct organism (form) emerges. Invariant proteins that I consider very likely candidates for making the design codes are among the growth factors, growth-factor receptors, proteins that are involved in signal transmission, proteins that are in control of DNA molecules, proteins that determine the polarity, segmentation, and symmetries in developing embryos.

A primordial cell can then develop into an organism. Its design code determines the design and the layout of the structures of the organism during its embryonic development.

Designing of forms are not bound to a certain point in time in the history of living organisms on earth. It is a process by which forms have been designed and emerged on a continuous basis. In the absence of urprimordial cells, other cells that are equivalent to them are being used in this process. Changes of DNA sequences, others than those that encode the invariant proteins of the design codes have no effect on the design codes of biological forms. Variable proteins, as the sequence analyses of these molecules from different organisms show, do not seem to be responsible for differences in designs of living organisms, or offer any indication for phylogenetic relations between them. On the contrary, those data can be interpreted in favor of my notion that the designs of biological forms cannot be touched by the changes of variable proteins. The organisms' subordinate regulative systems change and use the variable proteins according to the needs of the organism dictated by their design codes. The following examples cannot be interpreted otherwise: First, in a number of studies conducted in invertebrates and vertebrates similarities with regard to subordinate regulative systems between them during their development are evident. Second, studies are indicating that the subordinate regulative systems in different organisms are using similar structural proteins for building different organs and structures. For example, in Drosophila the same structural proteins are used for the formation of antennae and legs, or halters and wings. In the same organism in experiments where the loss of function of some homeotic proteins results in changes of thoracic parts of the body into structures appropriate to abdominal segments, and vice versa, it has been shown that both visceral and thoracic organs can be built from the same structural proteins. Third, the same structural proteins can be used by different organisms. The difference between the formation of a limb of a land vertebrate and the fin of a fish does not require that their structural proteins should be different from each other.

To illustrate the relation between the design code, subordinate regulative system, and structural proteins I compare the subordinate regulative systems in living organisms to a team of workers—masons, carpenters, painters, cement mixers, etc.—who are working together on construction sites with different designs. Those workers' duties are well-defined: they have to perform their jobs within the framework of the design of each construction. So the job of subordinate regulative systems in living organisms, indeed, are to work along the path of development of that organisms according to the instructions provided by their design codes. Concerning the use of the same structural proteins for different structures, I compare those proteins to the same construction material that the workers are using at different parts within a certain construction site. In a similar way the subordinate regulative systems use the structural proteins in different parts of an organism.

II. A reprogramming of a design code can lead to turning of a differentiated cell into an undifferentiated cell. Such cells, however, can reprimordialize, that means their invariant proteins can build a new design code for the emergence of a new form.

III. The proteins in design codes cannot be altered by environmental factors. If the amino acid sequence of any of the participating proteins is altered, the design code will obliterate. In case such an alteration happens inside an egg cell, one of the two possibilities is expected to occur. Either the affected cell is no longer viable, or develops into a body with an indistinct body design (malformed). If alteration of any of the proteins involved in a design code occurs in a cell of a developing or developed organism, the affected cell turns into a cancer cell. According to this notion, cancer gets a new definition, other than the one that is currently in use. In the relevant literature cancer has been considered as a result of accumulation of mutations in susceptible cells. Using a different phrase, cancer has been seen as clonal proliferation that arises owing to mutations that confer selective growth advantage on cells. However, we have to keep in mind that 'accumulation of mutations' and 'mutations that confer selective growth advantage on cells' are describing the changes being observed in cancer cells, but do not say anything about how any one of those mutations initiate cancer, i.e. turning a normal cell into a cancer cell. Screening for mutated genes in cancer cells lead to finding of more than 350 such genes so far. One may select the most significant ones of those genes and tries step by step to find out the kind of gene or genes that are supposedly offering growth advantage to a cell, and how it woks. The chances are that no mutation with the ability of conferring growth advantage will be found, unless one of the following two situations are given: a) The existence of a cancer making mechanism inside the cells that can be *turned on* by some mutations; this is very unlikely, because it is not in the interest of living organisms to carry such a potential of self-destruction within their cells. b) Mutations that are supposed to offer growth advantage have to be in position to change the whole chemistry of the affected cell; this is also unthinkable. According to the design code theory, any protein that its mutation prompts a cell to grow out of control is part of its design code. Mutations of such proteins by destabilizing the design code of the cell turn it into a 'defective urprimordial cell'. These cells are unable to recognize the boundary of the form, thus grow out of control.

IV. In accordance with the design code theory, adaptability is redefined. For the sake of their survival, living organisms need to evaluate the unfavorable conditions and circumstances under which they are living, unless these are beyond their control such as natural disasters. The essence of this evaluation is to overcome the obstacles and shortages that endanger the means they need to thrive and survive. Since the environment is steadily changing, the threats are changing, too, in form and magnitude. Therefore, living organisms must be able to stand up to those challenges. They can do this through changing their behavior, their biochemistry and cell functions. As far as changes of behavior are concerned, they would very likely look around for a safer nesting place, or eventually move to somewhere else, for example. When this deems to be insufficient, then biochemical and functional changes take place, such as changing of their skin color, camouflage and others. If they fail to do so, they would die. These changes, however, are not dictated by the environment; in other words, biological organisms do not change passively or randomly as it has been assumed by the the supporters of the evolution theory. They rather change themselves from within, thanks to the intelligence of their protein molecules. Thus, adaptability is a potential within the living organisms, which helps them to overcome the effects of the environmental changes, in order to maintain the integrity of their designs.

NOTE

I published my theory of design code in my book titled 'Primordialization, the way new living organisms emerge' in the year 2000. At that time I wanted to attract the attention of scientists who are interested in biodiversity. I continued my effort by presenting my design code (primordialization) theory at a number of international meetings, such as SMBE +AGA Meeting 2000 (1), Evolution Meeting 2001 (2), and in addition to them at the Teheran University, Rasht University (Iran) and Goettingen University (Germany).

In the year 2006, I tried to explain the significance of my theory with regard to the initiation of cancer at a Cold spring harbor laboratory meeting (3). After that in order to emphasis the significance of my theory with regard to this issue, I revised my book and changed its title to 'The design of biological forms, development, and initiation of cancer'. I discussed on this topic further at an EMBO conference in 2009 (4).

In this present version, I combined the first two chapters of the book, for simplicity reason, into one. I hope, with the choice of the new title I attract the attention of protein-sequence analysts, who by using their expertise could provide a list of primordial proteins.

1) 'Primordialization: A DNA-independent mechanism of diversity of living organisms', at the SMBA/AGA meeting (June 17-20, 2000), Yale University, USA.

2) 'Primordialization: The process by which invariant proteins determine the diversity of living organisms', at the Evolution 2000 meeting (June 26-July 1, 2001), University of Tennessee, USA.

3) 'A novel mechanism of cancerogenesis: Destabilization of cell's design regulative system', at the meeting on mechanisms and models of cancer (August 16-20, 2006), Cold spring harbor laboratory, USA.

4) 'Mutations of design regulative proteins during morphogenesis lead to initiation of cancer', at the conference on morphogenesis and dynamics of multicellular organisms (October2-6,2009), EMBO, Germany

REFERENCES

Aguinaldo, A. M. A. et al. (1997). Evidence for a clade of nematodes, arthropods and other molting animals. Nature 387, 489–493.

Akam, M. (1989). Hox and HOM: homologous gene clusters in insects and vertebrates. Cell 57, 347–349.

Alberts, B., Bray, D., Lewis, J., Raff, M., Robert, K. and Watson, J. D. (1989). Molecular Biology of the Cell, 2nd ed., Gerland Publishing, New York and London.

Alcock, J. (1989). Animal Behavior, An Evolutionary Approach, 4th ed., Sinauer Associates, Sunderland, Mass.

Anderson, K. V. (1987). Dorsal-ventral embryonic pattern genes of Drosophila. Trends in Genetics. 3, 91–97.

Andersson, D. I., Slechta, E. S. and Roth, J. R. (1998). Evidence that gene amplification underlies adaptive mutability of the bacterial lac operon. Science 282, 1133–1135.

Averof, M. (1998). Origin of the spider's head. Nature 395, 436-437.

Baulcombe, D. C. (1996). RNA as a target and an initiator of post-transcriptional gene silencing in transgenic plants. Plant. Molec. Biol. 32, 79-88.

Baulcombe, D. (2007). Amplified silencing. Science 315, 197-200.

Benton, M. J. (1990). Phylogeny of the major tetrapod groups: morphological data and divergence dates. Journal of Molecular Evolution 30, 409-424.

Benton, M. J. (2000). Stems, nodes, crown clades, and rand-free list: is Linnaeus dead? Biological Reviews 75, 633-648.

Beukeboom, L. W. et al. (2007). Haploid females in the parasitic wasp Nasonia vitripennis, Science 315, 206.

Blackburn, D. G. (1999). Placenta and placental analogs in reptiles and amphibians. In Encyclopedia of Reproduction (eds. E. Knobil, and J. D. Neil), pp. 840-847. Academic Press, San Diego.

Blackwelder, R. E. and Garoian, G. S. (1986). Handbook of Animal Diversity. CRC Press, Boca Raton, FL.

Bray, D., Levin, M. D. and Morton-Firth, C. J. (1998). Receptor clustering as a cellular mechanism to control sensitivity. Nature 393, 85-88.

Britten, R. J. and Kohne, D. E. (1968). Repeated sequences in DNA. Science 161, 529-540

Brooker, R. J. (1999). Genetic Analysis and Principles. Benjamin Cummings, California.

Camin, J. H. and Sokal, R. R. (1965). A method for deducing branching sequences in phylogeny. Evolution 19, 311-326.

Campos-Ortega, J. A. and Hartenstein, V. (1985). The Embryonic Development of Drosophila melanogaster. Springer, Berlin.

Casares, F. and Man, R. S. (1998). Control of antennal versus leg development in Drosophila. Nature 392, 723-728.

Cheng, T. C. (1964). The Biology of Animal Parasites. Saunders, Philadelphia.

Chirpich, T. P. (1975). Rates of protein evolution: a function of amino acid composition. Science 188, 1022-1023.

Coyne, J. A. and Orr, H. A. (2004). Speciation. Sinauer Associates, Sunderland, Mass.

Darwin, C. R. (1859). On the Origin of Species by Means of

Natural Selection. John Murry, London.

DeDuve, C. (1991). Blueprint for a Cell: The Nature and Origin of Life. Neil Patterson, Burlington, NC.

Dell, H. (2007). Marked from the start. Nature 445, 157.

DiBerardino, M. A., Orr, N. H. and McKinnell, R.G. (1986). Feeding tadpoles cloned from Rana erythrocyte nuclei. Proceeding of the National Academy of Sciences, USA 83, 8231-8234.

Dickerson, R. E. (1971). The structure of cytochrome c and the rates of molecular evolution. Journal of Molecular Evolution 1, 26-45.

Dogiel, V. A., Polyanski, Yu. I. and Kheisin, E. M. (1966). General Parasitology. Academic Press, New York.

Driever, W. and Nüsslein-Volhard, C. (1988). The bicoid protein determines position in the Drosophila embryo in a concentration dependent manner. Cell 54, 95-104.

Duboule, D. and Dolle, P. (1989). The structural and functional organization of the maurine Hox gene family resembles that of Drosophila homeotic genes. Journal of European Molecular Biology Organization 8, 1497-1505.

Engel, R. (1990). Mating-type genes, meiosis and sporulation.

In: Molecular Biology of the Fission Yeast (eds. A. Nasim, B. Johnson, and P. Young), pp. 32-37. Academic Press, San Diago.

Fitch, W. M. and Margoliash, E. (1967). Construction of phylogenetic trees. Science 155, 279-284.

Frankhauser, G. (1952). Nucleo-cytoplasmic relations in amphibian development. International Review of Cytology 1, 165-193.

Frankhauser, G. (1955). In: Analysis of Development (eds. B. H. Willer, P. A. Weiss, and V. Hamburger), pp. 126-150. Saunders, Philadelphia.

Gehring, W. J. and Nöthiger, R. (1973). The imaginal discs of Drosophila. In: Developmental Systems: Insects (eds. S. Counce and C. H. Waddington), Vol. 2, pp. 211-290, Academic Press, New York.

Gehring, W. J., Müller, M., Affolter, M., Percival-Smith, A., Billeter, M., Qian, Y. Q., Otting, G. and Wüthrich, K. (1990). The structure of the homeodomain and its functional implications. Trends in Genetics. 6, 323-329.

Gibbons, A. (1998). Which of our genes make us humans? Science 281, 1432-1434.

Gibbs, R. A. (2007). Evolutionary and biochemical insights from the rhesus macaque genome. Science 316, 222-234.

Gilbert, S. F. (1994). Developmental Biology, 4th ed., Sinauer Associates, Sunderland, Mass.

Goodman, M., Moore, G. W. and Matsuda, G. (1975). Darwinean evolution in the genealogy of heamoglobin. Nature 253, 603-608.

Graham, A., Papalulu, N. and Krumlauf, R. (1989). The maurine and Drosophila homeobox gene complexes have common features of organization and expression. Cell 57, 367-374.

Greenman, C. et al. (2007). Patterns of somatic mutation in human cancer genomes. Nature 446, 153-158.

Griffith, G. M., Berek, C., Kaartinen, M., and Milstein, C. (1984). Somatic mutation and the maturation of the immune response to 2-phenyl oxazolone. Nature 312, 271-275.

Gurdon, J. B. (1968). Transplanted nuclei and cell differentiation. Scientific American 219 (6), 24-35.

Hammord, S. M., Bernstein, E., Beach, D. and Hannon, G. I. (2000). An RNA-directed nuclease mediates post-transcriptional gene silencing in Drosophila cells. Nature 404, 293-296.

Hannon, G. J. (2002). RNA interferance. Nature 418, 244-251.

Harris, R. A., Rogers, J. and Milosavljevic, A. (2007). Human-specific changes of genomic structures detected by genomic triangulation. Science 316, 235-237.

Hiller, L. W. et al. (2004). Sequence and comparative analyses of the chicken genome provide unique perspective on vertebrate evolution. Nature 432, 695-716.

Hillis, D. M., Moritz, G. and Mable, B. K. (1996). Molecular Systematics. Sinauer Associates, Sunderland, Mass.

Hopkin, M. (2007). Chimps lead evolutionary race. Nature 446, 841.

Hoyer, B. H., McCarthy, B. J. and Bolton, E. T. (1964). A molecular approach in the systematics of higher organisms. Science 144, 959-967.

Ingham, P. W. (1988). The molecular genetics of embryonic pattern formation in Drosophila. Nature 335, 25-34.

Irion, U. and Johnston, D. St. (2007). Bicoid RNA localization requires specific binding of an endosomal sorting complex. Nature 445, 554-558.

Johns, B. and Miklos, G. L. G. (1988). The Eukaryote Genome in Development and Evolution. Allen and Unwin, London.

Jukes, T. H. and Holmquist, R. (1972). Estimation of

evolutionary changes in certain homologous polypeptide chains. Journal of Molecular Biology 64, 163-179.

Kay, R. F., Ross, C. and Williams, B. A. (1997). Anthropoid origins. Science 275, 797-804.

Kazemie, M. (2000). A DNA-independent mechanism of diversity of living organisms. Meeting of the Society for Molecular Biology and Evolution, New Haven, Conn.

Kazemie, M. (2001). Primordialization: The way new living organisms emerge. Universal Publishers, USA.

Kazemie, M. (2001). Invariant proteins determine the designs of the biological forms. Meeting of the Society for the Study of Evolution, Knoxville, TN.

Kazemie, M. (2006). A new mechanism of cancerogenesis: Destabilization of cell's design regulative system. Cold Spring Harbor Meeting on Mechanisms and Models of Cancer. Cold Spring Harbor, NY.

Kazemie. M. (2009). The designs of biological forms, development, and initiation of cancer. Author House, USA.

Kazemie, M. (2009). Mutations of design regulative proteins during morphogenesis lead to initiation of cancer. EMBO Conference on Morphogenesis and Dynamics of Multicellular Systems. EMBL, Heidelberg.

Kimura, M. (1968). Evolutionary rate at the molecular level. Nature 217, 624-626.

King, J. L. and Jukes, T. H. (1969). Non-Darwinian Evolution. Science 164, 788-798.

King, M-C. and Wilson, A. C. (1975). Evolution at two levels in humans and chimpanzees. Science 188, 107-116.

King, N., Hittinger C. T. and Carroll S. B. (2003). Evolution of key cell signalling and adhesion protein families predates animal origin. Science 301, 361-363.

Klobutcher, L. A. and Prescott, D. M. (1986). The special case of the hypotrichs. In: The Molecular Biology of Ciliated Protozoa (ed. J. A. Gall), pp. 111-154. Academic Press, Orlando, FL.

Klug, W. S. and Cummings, M. R. (1997). Concepts of Genetics, 5th ed., Printice Hall, Upper Saddle River, NJ.

Konigsberg, T. R. (1986). The embryonic origin of muscle. In : Myology (eds. A. Engel and B. Q. Banker), Vol. 1, pp. 39-71. McGraw-Hill, New York.

Kuzloff, E. N. (1990). Invertebrates. Saunders College Publishing, Philadelphia.

Kumar, S. and Hedges, B. (1998). A molecular timescale for vertebrate evolution. Nature 392, 917-920.

Langley, C. H. and Fitch, W. M. (1974). An examination of the constancy of the rate of molecular evolution. Journal of Molecular Evolution 3, 161-177.

Levin, M., Johnson, R. L., Stern, C. D., Kühn, M. and Tabin, C. (1995). A molecular pathway determining left-right asymetry in chick embryogenesis. Cell 82, 803-814.

Loomis, W. F. (1975). *Dictyostelium discoideum*. A Developmental System. Academic Press, New York.

Luo, Zhe-Xi, Li, P. G. and Chen, M. (2007). A new eutriconodont mammal and evolutionary development in early mammals. Nature 446, 288-293.

Mallet, J. (2007). Hybrid speciation. Nature 446, 279-283.

Manning, J. E., Schmidt, C. W. and Davidson, N. (1975). Interspersion of repetitive and nonrepetitive DNA sequences in the *Drosophila melanogaster* genome. Cell 4, 141-155.

Marris, E. (2007). The species and the spacious. Nature 446.

Mayr, E. and Ashlock, P. D. (1991). Principles of Systematic Zoology, 2nd ed., McGraw-Hill, New York.

McKenna, M. C. (1975). Towards a phylogenic classification of the mammalia. In: Phylogeny of the Primates: A

Multideciplinary Approach (eds. W. P. Luckett and F. S. Szolay), pp. 21–44. Pleneum Press, New York.

Morris, H., Taylor, G., Masento, M., Jermin, K. and Kay, R. (1987). Chemical structure of the morphogen differntiation inducing factor from *Dictyostelium discoideum*. Nature 328, 811-814.

Müller, W. A. (1997). Developmental Biology. Spinger-Verlag, New York.

Nickoloff, J. A. and Hoekstra, M. F. (1998). DNA Damage and Repair. Humana Press, Totowa, NJ.

Nomura, M. and Li, E. (1998). Smad2 role in mesoderm formation, left-right patterning and craniofacial development. Nature 393, 786-790.

Novacek, M. J. (1992). Mammalian Phylogeny: shaking the tree. Nature 356, 121-125.

Nüsslein-Volhard, C., Frohöfer, H. G. and Lehman, R. (1987). Determination of antero-posterior polarity in Drosophila. Science 238, 1675-1681.

Organ, C. L., Shedlock, A. M., Meade, A., Pagel, M. and Edwards, S. V. (2007). Origin of avian genome size and structure in non-avian dinosaurs. Nature 446, 180-184.

Pennisi, E. (1998). How the genome readies itself for

evolution. Science 281, 1131–1134.

Rajewsky, K., Forster, I. and Cumano, A. (1987). Evolutionary and somatic selection of the antibody repertoire in the mouse. Science 238, 1088–1094.

Reddy, S. K., Rape, M., Margansky, W. A. and Kirschner, M. W. (2007). Ubiquitination by the anaphase–promoting complex drives spindle checkpoint inactivation. Nature 446, 921–925.

Rhesus Macaque Genome Sequencing and Analysis Consortium. (2005). Chimp genome sequence. Nature 437, 69–87.

Riggs, A. (1959). Molecular adaptation in haemoglobins: nature of the bohr effect. Nature 183, 1037–1038.

Ryan, A. K., Blumberg, B., Rodriguez-Esteban, C., Yonei-Tamura, S., Tamura, K., Tsukui, T., de la Pena, J., Sabbag, W., Greenwalk, J., Choe, S., Norris, D. P., Robertson, E. J., Evans, R. M., Rosenfeld, M. G. and Izpisua Belmonte, C. (1998). Pitx2 determines left–right asymmetry of internal organs in vertebrates. Nature 394, 545–551.

Schneiderman, H. A. (1976). In: Insect Development (ed. P. A. Lawrence), pp. 3–34, Blackwell, Oxford, UK.

Shephard, J. C. W., McGinnis, W., Carrasco, A. E., De Robertis, E. M. and Gehring, W. J. (1984). Fly and frog homeodomains show homologies with yeast mating-type regulatory proteins. Nature 310, 70-71.

Shinde, U. P., Liu, J. J and Inouye, M. (1997). Protein memory through altered folding mediated by intramolecular chaperones. Nature 389, 520-522.

Simpson, G. G. (1964). Organisms and Molecules in Evolution. Science 146, 1535-1538.

Sjoeblem, T. et al. (2006). The consensus coding of sequences of human breast and colorectal cancers. Science 314 268-274.

Smith, J. C. (1989). Mesoderm induction and mesoderm-inducing factors in early amphibian development. Development 105, 665-677.

Sokal, S., Wong, G. G. and Melton, D. (1990). A mouse macrophage factor induces head structures and organizes a body axis in Xenopus. Science 249, 561-564.

Stegmeier, F. et al. (2007). Anaphase initiation is regulated by antagonistic ubiquitination and deubiquitination activities. Nature 446, 876-881.

Sternberg, P. W. and Horwitz, H. R. (1986). Pattern formation during vulva development in *C. elegans*. Cell 44, 761-772.

Thompson, J. A, Itskovitz-Eldor, J., Shapiro, S. S., Waknitz, M.

A., Swiergiel, J. J., Marshall, V. S. and Jones, J. M. (1998). Embryonic stem cell line derived from human blastocytes. Science 282, 1145-1147.

Torres–Pallida, M-E., Parfitt, D-E., Kouzarides, T. and Zernicka-Goetz, M. (2007). Histone argenine methylation regulates pleuripotency in the early mouse embryo. Nature 445, 214-218.

Vogel, G. (1998). A two–piece protein assembles itself. Science 281, 763-764.

Wang, J. et al. (2007). Opposing LSD1 complexes function in developmental gene activation and repression programs. Nature 446, 882-887.

Webster, G. and Goodwin, B. (1996). Form and Transformation, Generative and Relational Principles in Biology. Cambridge University Press, New York.

Wilkins, A. S. (1993). Genetic Analysis of Animal Development, 2nd ed., Wiley-Liss, New York.

Wellik, D. M. and Capechi, M. R. (2003). Hox 10 and Hox 11 genes are required to globally pattern the mammalian skeleton. Science 301, 363-367.

Whilfield, J. (2007). We are family. Nature 446, 247-249.

Wilson, A. C., Sarich, V. M. and Maxon, L. R. (1974). The importance of gene rearrangements in evolution: evidence from studies on rates of chomosomal, protein, and anatomical evolution. Proceedings of the National Academy of Sciences, USA 71, 3028-3030.

Zuckerkandl, E., Derancourt, J. and Vogel, H. (1971). Multitional trends and random process in the evolution of informational macromolecules. Journal of Molecular Biology 59, 473-490.

www.ingramcontent.com/pod-product-compliance
Lightning Source LLC
Chambersburg PA
CBHW022103170526
45157CB00004B/1461